Funny Bumble was pretty but pesky.

"Boo!" Funny said to Daddy Longlegs.

"Boo!" Funny yelled to Teensy Tick.

Queen Bug was resting.
"Boo!" Funny yelled.

"You pesky bee!" she said.
"That was not funny!"

"You will buzz as you fly! Go!"

Buzz, buzz, buzz went Funny's wings.

And bumblebees still buzz today.